## OF
# WHITE-TAILED DEER

## BY DAVE AND STEVE SHELLHAAS

# Outdoor IQ

**OutdoorIQ's ABC Books of Hunting Heritage and Habitat**
published by

Miami Valley Outdoor Media
P.O. Box 35
Greenville, OH  45331

www.OutdoorIQ.org

The Hunting Heritage and Habitat Book Series include a collection of books published by OutdoorIQ to support the various conservation organizations in their efforts to promote habitat conservation and youth hunting programs.  Proceeds from the sale of each book will be donated back to the partnering conservation organization to support their habitat and youth programs.

Photo Credits
iStock photos

ISBN 978-0-9845251-8-8

#  **is for antler.**

Antlers are made of bone and come in many shapes and sizes. The antlers on a buck's head grow back each year.

# B is for buck.

Male deer are called bucks. White-tailed bucks usually have much bigger bodies than female deer.

# C is for camouflage.

The brown hair of the deer helps them to be camouflaged and hide in the forest. Can you see how the color of the deer matches the color of its surroundings?

# D is for doe.

A female deer is called a doe. Does spend most of their time with their fawns during the first year.

# E is for ears.

Deer have very large ears that can move in different directions. These big ears allow them to hear very well and know when danger is coming.

# F is for fawn.

Baby deer are called fawns. Fawns have spots that help them hide in the tall grass.

# G is for grunt.

A buck deer makes a sound in the fall to let other bucks and the does know he is there. This sound is called a grunt. Grunts can be loud or they can be soft.

# H is for herbivore.

Deer are herbivores. Herbivores are animals that only eat plants. Deer like to eat grass, clover, acorns, and many other plants.

# I is for insulated.

Deer hair is very thick and hollow to insulate the deer from the cold. The deer's hair is so insulated, snow will actually stay on their back and will not melt.

# J is for jump.

Deer are great jumpers. Deer can jump over fences that can be up to six or seven feet high.

# K is for kick.

Kicking is a way deer protect themselves. They also scare other deer away by kicking them.

# L is for licking branch.

Bucks mark their territory by pawing the ground to make what is called a scrape. There is usually a licking branch above the scrape that the buck licks, chews, and rubs over his face. This leaves his scent for other deer to smell.

# M is for May.

May is the month that many fawns are born. This time of year is usually warmer and there is grass and plants to hide the fawns.

# N is for nontypical.

Most antlers of white-tailed deer have points that grow upward and match on both sides. Nontypical antlers are antlers that have points growing in all directions and do not match.

# O is for offspring.

Offspring is the word used to describe the babies of animals. The offspring of the white-tailed deer are their fawns.

# P is for points.

Bucks grow antlers with points. Buck deer are described by how many points they have. Can you count the points on this buck's antlers?

# Q is for quiet.

Fawns stay very quiet and still as they hide in the grass or tall plants. This keeps the fawns from being found and keeps them safe from danger.

# R is for rub.

Some trees in the forest may look like this. This is called a rub because bucks use their antlers to rub the bark off of the trees. Rubs are like signs telling other bucks and does that the buck was there.

# S is for shed antler.

Shed antlers are antlers that fall off the buck's head in late winter. Bucks shed their antlers every year. The antlers then grow back in the spring and summer.

# **T** is for typical.

Typical antlers have points that grow upward and are matching on both sides. Do you see how this buck's antlers match on both sides?

# U is for urban.

Urban means living in or near a city or town. Many deer live in urban areas near people's houses.

# **V** is for velvet.

Velvet is the soft, fuzzy covering over the buck's antlers when they are growing over the summer. In the fall, the velvet dries as the antlers hardens. The buck then rubs the velvet off on trees.

# W is for whitetails.

White-tailed deer are also called whitetails and get their name from their large white tail. When the deer get scared or excited, they show the white side of their tail like a flag.

#  **X** is for 4 x 4.

In some places, people describe bucks by how many points they have on each side. If a buck has four points on each side, people say the buck is a 4 X 4.

# Y is for yarding up.

Yarding up is when deer come together in large groups. Deer yard up in the winter when the weather is cold and snow is deep.

# **Z** is for zero.

Zero is the number of white-tailed deer there will be if we do not conserve important habitat and manage their populations. Whitetails Unlimited is very important is conserving and increasing habitat and helping to educate the public about conservation measures.

# Whitetails Unlimited

Since our beginning in 1982, Whitetails Unlimited has remained true to its mission and has made great strides in the field of conservation. We have gained the reputation of being the nation's premier organization dedicating its resources to the betterment of the white-tailed deer and its environment. Our mission is to preserve the hunting tradition for future generations; promote the acquisition, enhancement, and management of wildlife habitat; and develop and promote educational programs related to wildlife conservation and the shooting sports.

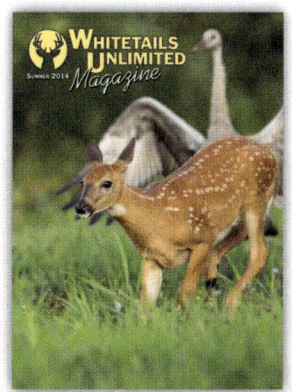

The white-tailed deer (Odocoileus virginianus) ranks second to none when it comes to research, popularity, management, hunting opportunity, and population. Through the combined efforts of state wildlife agencies and national conservation groups like Whitetails Unlimited, current estimates place North America's whitetail population at over 30 million, including subspecies, currently distributed throughout the continental United States and Canada.

At Whitetails Unlimited we see a bright future for the whitetail and work daily to make sure it remains that way. Without a doubt, whitetails are the most adaptable, widely distributed, and to us, one of the most fascinating animals on the face of the earth. Whitetails Unlimited is proud to proclaim that at no period in American history have whitetails been healthier, more numerous, nor have their future prospects been brighter.

# Habitat, Hunting Heritage and Wildlife Conservation

These three things are very closely connected in America. Not only are conservation organizations important to habitat and wildlife conservation, hunters and sportsmen also play an important role in conservation efforts. In North America, hunting is a very important part of raising the money necessary to conserve and protect habitat for all types of wildlife.

There is a 10 to 12 percent tax on hunting and shooting equipment that goes directly to fund conservation efforts in this country. This system was set up by the government in 1937 as a way to help fund important wildlife and habitat conservation efforts. Since it was begun in 1937, hunters and sportsmen have contributed over 10 billion dollars toward wildlife conservation.

This is why our hunting heritage is so important. Without hunters, this money would not be raised and the money available for wildlife conservation would be much less. The hunting heritage in this country is not only a wonderful family tradition; it is essential to the wildlife that calls our wonderful country home. We must ensure that our hunting heritage is passed to the next generation so that wildlife conservation can continue to be funded.

There is no alternative funding system to replace the potential lost funds for conservation. If hunting ended, most funding for wildlife conservation would be lost. This is why OutdoorIQ has partnered with conservation organizations and created these wonderful ABC Books for Hunting Heritage and Habitat. A portion of each book purchased will help these great organizations continue to fund youth initiatives and introduce the next generation to hunting.

Wildlife conservation in this country is dependent on the continuation of our hunting heritage. By purchasing this book for your child, you have been a part of this important effort to protect our habitat, wildlife, and hunting heritage.

# ABC
# BOOKS
### OF HUNTING HERITAGE AND HABITAT

This is one of several ABC books available to teach children not only the letters of the ABCs, but also about the ABCs of habitat and hunting. These books are written and published by OutdoorIQ who specializes in creating educational children's books about the outdoors. OutdoorIQ is committed to educating children and youth about the outdoors and the wonderful activities we can all share.

Conservation organizations such as Whitetails Unlimited are important to protecting the habitat necessary for the survival and well-being of all types of wildlife. These organizations are also committed to passing on the heritage of hunting and the outdoors to our children and youth.

For this reason, OutdoorIQ has partnered with these wonderful conservation organizations to create this ABC Books for Hunting Heritage and Habitat book series. Proceeds from the purchase of each book associated with the respective organization, goes back to the organization to help fund their youth programs and protect habitat.

By purchasing this book for your child, you have not only invested in your child's learning, you have also contributed to a worthy organization who is dedicated to protecting habitat and our hunting heritage. Thanks for your investment in the future of your child and future of our wildlife and hunting heritage.

If you and your child enjoyed this ABC book, check out the ABC books about the other wildlife and conservation groups. There are also other educational outdoor books available for older children and youth at the OutdoorIQ's website, www.OutdoorIQ.org.

OutdoorIQ

Made in the USA
Middletown, DE
20 December 2022